SEE, MAKE AND DO

AT THE SEASHORE

PAMELA HICKMAN
Illustrations by Twila Robar-DeCoste

Formac Publishing Company Limited
Halifax

Copyright © 1996 by Pamela Hickman. All rights reserved. No part of this book may be reproduced or transmitted in any form or by any means, electronic or mechanical, including photocopying, or by any information storage or retrieval system, without permission in writing from the publisher.

Canadian Cataloguing in Publication Data

Hickman, Pamela
  At the seashore
  Includes bibliographical references.

ISBN 0-88780-356-3

1. Seashore biology — Juvenile literature. 2. Nature study — Activity programs — Juvenile literature. I. Robar-DeCoste, Twila L. (Twila Lee), 1956- . II. Title

QH95.7.H54 1996 j574.5æ2638 C96-950022-X

Formac Publishing Company Limited
5502 Atlantic Street
Halifax, Nova Scotia B3H 1G4

Acknowledgements:
I wish to thank the following people for their assistance with my research for this book: Ken Adams, Curator of the Fundy Geological Museum; Bob Ogilvie, Curator of Special Places at the Nova Scotia Museum; Bob Pietrzak, Canadian Hydrographic Services; and Dr. Cheryl Lycette, N.D. for her advice on jellyfish stings. I would also like to acknowledge Dr. Merritt Gibson's excellent field guide, *Seashores*. Finally, I'd like to thank my friend Anne Kehoe for many enthusiastic seashore explorations and terrific campfires on the beach.

Dedication:
I would like to dedicate this book to the memory of David Threlford who shared his knowledge, enthusiasm and love of nature with so many.

Contents

Welcome to the seashore 4
Heading out 6
Tide talk 8
From sand to sea 10
Gearing up 12
Beach "buddies" you can make 14

On the beach 16
A beach is born 18
Sand tracks and trails 20
Dig it 22
Shell collecting 24
Beachcombing 26
The seashore at night 28

Rocky shores 30
On the rocks 32
Exploring a rock pool 34
Seaweed secrets 36
Living under the surface 38
Finding fossils 40
Campfires at the shore 42

Some shoreline animals up close 44
Wings over water 46
Shorebirds 48
Jellyfish 50
What a crab! 52
Saltmarshes 54
Become a nature detective 56
Animals ahoy! 58
S.O.S Save our shorelines 60

East coast national, provincial and state parks 62
Suggested field guides 63
Index 64

Welcome to the seashore

Whether you are heading to your local beach or travelling to shores unknown, this book will give you lots of ideas of what to take, things to do, what to look for and where to look. In this section you'll find out about tides and learn how to make your own tide table. Discover the different life zones on a beach and make some great "buddies" to help you dig down for a closer look at shore life.

Heading out

An outing to the seashore is a great way to have fun and meet some of nature's most fascinating creatures. Whether you dig in the mud, poke into rock pools, paddle in the shallows or just lie on the beach, this book will help you and your family discover many new adventures that await you at the shore.

To ensure that your shoreline activities are safe for you, and for the creatures that live there, here are a few safety and conservation tips to keep in mind.

SAFETY TIPS

1. Children should explore in groups of two or more. Adult supervision is recommended.

2. Check a tide table so you know when high tide is expected. Certain shorelines, particularly along the Bay of Fundy and Minas Basin, have extremely high tides that come in very quickly. The best time to explore the shorelines below the cliffs in these areas is after high tide, when the tide is going out.

3. Where possible, avoid walking on wet and algae-covered rocks — they are very slippery.

4. Take time to check out the shoreline before planning a long walk. Shoes are recommended for walking on barnacle-covered rocks. Tidal flats are often very muddy. As shoes and boots can get stuck in the mud, go barefoot.

5. Avoid rocky shorelines during high winds and storms. Large, powerful waves can appear without warning.

CONSERVATION TIPS

1. Always take your garbage away with you. Pack an extra plastic bag so you can pick up garbage left by others and make the

shoreline a better place for everyone. Join up with friends and neighbours for the annual Beach Sweep and help clean up the garbage on a certain stretch of shoreline during Environment Week, the first week in June every year.

2. Handle living creatures carefully. When you are finished looking at them, always return them to where they were found.

3. If you find a nest, watch it from a safe distance and make sure that the adult animals are not frightened away. Do not touch the nest, eggs or young.

4. Enjoy the wildflowers where they grow, instead of picking them.

5. Do not climb on sand dunes. They are very sensitive to disturbance and can be easily damaged by trampling (see p. 11 for more on sand dunes).

6. It is illegal to dig and remove fossils from the shoreline cliffs and reefs. For more on fossil finds, see p. 40.

Tide schedules

Check the local tourist bureau or newspaper for the current tide times in your area (the tide times are specific to each section of shoreline). For information on purchasing a set of tide tables for Atlantic Canada, contact the Canadian Hydrographic Service in Halifax. Or, turn the page and find out how to make your own tide table.

Tide talk

Take a walk on the ocean floor. Impossible? When you walk out on the mud flats, you are really walking on part of the ocean floor. Twice a day the ocean waters flow away from the shore; this is low tide. When the tide returns, it floods the mud with one to sixteen metres of water, depending which part of the coast you are on. The mudflat that you walked is now deep underwater.

What causes the tides? Tides are created by the gravitational pull of the Moon and Sun. Since the Moon is much closer to Earth than the Sun, its pull is stronger. Twice a month, when the Earth, Moon and Sun are in a straight line, the pull is strongest and the tides are highest. This occurs at the New Moon and the Full Moon and these tides are called spring tides. When the Moon, Earth and Sun make three points of a triangle, the tides are lowest, called neap tides. The height of tides is also affected by the shape of the ocean floor and the coastline.

TELLING THE TIDE

When you go down to the shore, look for the high tide mark along the sand or rocks. You can usually see a line of seaweed, driftwood and other things that have been left high and dry by the water. On large rocks, you may be able to see a difference in the kinds of plants and animals that are attached above and below the high tide mark.

Make a tide calendar

There are two tides per day on the Atlantic shore. It is a bit longer than six hours between high tide and low tide. You can calculate the tides for a shore near you with some simple math. Then you'll be able to predict high tide and low tide all summer, or even longer.

Remember, the time of high tide at one part of the shore can be very different from high tide at another place some distance along the coast.

You'll need:
- a calendar
- a current tide schedule
- two different coloured pencils

- Check your local newspaper for the high and low tide times at a shore near you. Mark these times, for the specific date they are given, on your calendar. Use one colour for high tides and the other for low tides.

- To figure out when high and low tide will be from one day to the next, add 50 minutes to the time of the tide. For example, if high tide comes at 6:00 a.m. and 6:25 p.m. on one day, it will come again at approximately 6:50 a.m. and 7:15 p.m. the next day. The same is true for low tides. By doing these simple calculations for each day on your calendar, you can make your own tide table for your section of the shore.

From sand to sea

As you explore the beach, you'll discover that the kinds of plants and animals change as you move from the dry sand dunes to the water. Each species does best in a certain area, or zone, of the beach. Some species prefer the driest zone of the dunes and upper beach (zone A), while others are adapted to the tides and can survive regular flooding (zone B). Certain plants and animals need to be underwater at all times (zone C). When you visit a beach, try to find the main beach zones illustrated below.

American Beachgrass

Seaside Plantain

Sea Rocket

Zone A

Zone B

Help protect the dunes

When strong winds blow sand up the beach over a period of time, sand dunes are created. The wind keeps blowing the sand back from the shore until it is trapped by growing plants such as American Beach-grass. Its roots hold onto the sand and keep it from blowing away. As new plants grow, the dune becomes more stable and insects and other small animals move in. The entire habitat depends on keeping the sand from blowing away. For this reason, it is very important that people stay off the dunes. Climbing on sand dunes damages the plants and exposes their roots so they can't hold onto the sand any more. When the sand is left unprotected, the wind will again blow it away. Other plants will be smothered by blowing sand and the dune habitat can be destroyed by a "blow out."

Edible Kelp

Zone C

Eelgrass

Gearing up

Before you head to the shore with your family or friends, pack your knapsack with some "beach buddies" and dress comfortably. Here are some tips on what to take.

BEACH BUDDIES
- snacks or lunch and fresh water
- sunscreen
- insect repellant (on calm days mosquitoes, blackflies, deerflies and horseflies can be a nuisance)
- first aid kit
- matches
- a towel
- an extra jacket or sweater
- binoculars
- field guides, such as birds, insects, seashells, seashores (see list on page 63)
- a plastic magnifying glass
- pad and pencil for making notes or sketches
- a camera
- something to dig with in the mud (see page 14)

- an underwater viewer (see page 15)
- a container for a temporary aquarium (see page 22)
- some plastic bags for collecting "treasures" and garbage
- picnic blanket
- a hat to keep off sun and insects
- loose-fitting pants that can be rolled up
- old sneakers for walking on rocks
- knapsack
- bathing suit
- a watch

Beach "buddies"

A super scooper

Digging in the sand and mud is lots of fun. It's also a great way to find some of the creatures that live there. Here's a super scooper you can make out of an old plastic jug.

You'll need:
- a rinsed out 2-L juice jug with handle and lid (old bleach containers work well too)
- scissors
- sandpaper

1. With the handle facing up, ask an adult to cut out the lower part of the container, as shown.

2. Sand the cut edges until they are smooth.

you can make

An underwater window

Take a peak under water with this easy-to-make viewer. You'll need:
- a rinsed out 1-L juice can
- a can opener
- waterproof tape (duct tape works well)
- clear plastic wrap

3. Stretch a piece of plastic wrap across one end. Tape the edges securely to the can.

1. Use the can opener to take both ends off of the can.

2. Tape the sharp edges of the can so you don't cut yourself.

4. To use the viewer, place the end with the plastic wrap in the water. Do not let the water come up over the open end. Look into the open end and see what's happening underwater.

On the beach

Find out where beaches come from and then discover what lives on and under the sandy shore. Dig up some of the hidden creatures in the mud so you can watch them more closely in a temporary home that you make. Comb the shores for "treasures" and make some wind chimes out of shells. Then come back to the shore at night and discover a whole new world of creatures that come out in the dark.

A beach is born

Have you ever wondered where all of the sand on a beach comes from? Follow the scenes below to see how a beach is born.

1. The grains of sand start out as part of much larger rocks, perhaps a long way from the shore.

2. Wind and water are constantly eroding, or wearing away, the rocks. The tiny particles are carried to the sea by rivers and are dumped on the ocean floor.

3. Currents and tides carry the sand up from the bottom and deposit it on the shore.

4. Waves move the sand higher onto the land, creating a beach and wind blows the sand farther up the beach, forming sand dunes.

Sand crafts

Sand is not only terrific to walk on and dig in, but you can also use it to make some great crafts. Try out some of these ideas.

- Draw a simple pattern on a piece of stiff paper or thin cardboard. Go over your lines with a thin layer of white glue and then sprinkle sand over the paper. The sand will stick to the glue and a sand picture will appear.

- Place some sand in two or three different plastic containers (yoghurt or margarine containers work well). Throroughly mix a different colour of powdered tempera paint into each container. Scoop layers of different coloured sand into a tall glass jar or vase to make a pretty anchor for dried flowers, candles, or whatever. By poking long sticks into the jar you can mix the colours in certain places and create patterns

- Mix some sand into liquid tempera paint to get a textured effect in your picture.

Sand tracks and

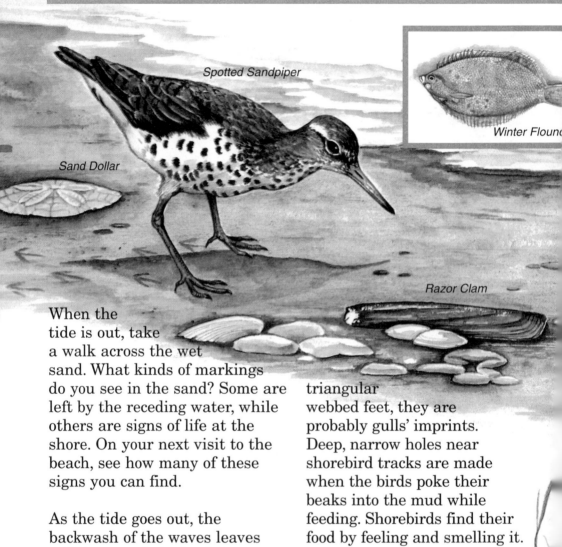

When the tide is out, take a walk across the wet sand. What kinds of markings do you see in the sand? Some are left by the receding water, while others are signs of life at the shore. On your next visit to the beach, see how many of these signs you can find.

As the tide goes out, the backwash of the waves leaves widely spaced ripples lying parallel to the shore.

Bird tracks are easy to spot. Narrow, three-toed tracks are left by small shorebirds such as sandpipers, sanderlings and plovers. If you see large, triangular webbed feet, they are probably gulls' imprints. Deep, narrow holes near shorebird tracks are made when the birds poke their beaks into the mud while feeding. Shorebirds find their food by feeling and smelling it.

Small, shallow depressions in the sand are signs of flounder. They come in with the tide and take "bites" out of the sand in their search for food in the mud.

trails

Lugworms live in the sand in the intertidal zone. With some practice you can find their tiny U-shaped burrows. The front opening is funnel-shaped and the narrower back opening is marked by tiny piles of castings, similar to the castings of earthworms.

If you notice water squirting out of groups of holes in the sand or mud as you approach, you have probably found a colony of Soft-shelled Clams. They are often found near a freshwater inlet. Razor Clams also squirt water and then quickly burrow deeper into the sand.

At low tide look for the outlines of Bar Clams buried just below the surface.

Dig it

Once you find signs that life is lurking beneath your toes in the wet sand, start digging. Use your home-made scooper (see page 14) or a trowel and dig down at one side of a creature's hole in the sand. Work quickly since some of these animals, especially razor clams, can rapidly dig deeper. When you find something, carefully transfer it to the "mobile home" described below. Handle clam worms and bloodworms carefully as they will nip.

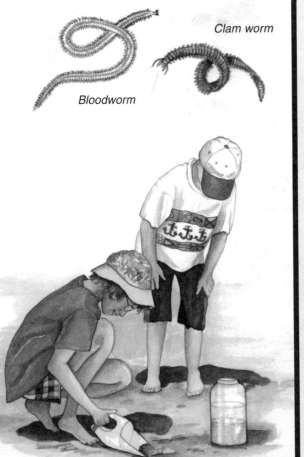

Clam worm

Bloodworm

Make a "mobile home"

Set up this temporary aquarium for a closer look at how the animals move and feed.

You'll need:
- a rinsed out 4-L jar (a large condiment jar works well)
- some sand and sea water
- a scoop
- shoreline creatures dug out of intertidal mud
- a piece of cheesecloth or fine screening
- an elastic band

1. Scoop some sand into the bottom of the jar until it is about half full.

2. Add some animals that you have dug up in the intertidal zone, such as clams, mussels, worms etc.

1. Gently pour some sea water in the jar until it is two thirds full.

2. Put a piece of cheesecloth or fine screening over the jar and secure it with an elastic band.

3. Place the jar in the shade and let the sand settle. Since the sand is covered with water, it is similar to high tide and the animals should be active. Watch for worms swimming around. Can you see a clam moving along on its fleshy foot? Try to find a clam's tube-like siphons sticking out between its shells. A siphon sucks water and nutrients into the shell where the microscopic food is trapped and eaten. Waste water is discarded through another siphon.

4. When you have finished watching the creatures, return them to where you found them.

CLAMMING

At low tide from April to October you may see people bent over and digging on the mud flats. What are they digging for? The answer is soft-shelled clams. In addition to the sea ducks, crabs, lobsters and flatfish that naturally feed on clams, people also join in the harvest. Whether you dig your own clams or buy them fresh (live) from a local market, try this simple method of preparing them at home or on a campfire.

1. Make sure the shells of fresh clams are tightly closed. Discard open shells since they mean that the clam is dead.

2. Scrub the shells under cold water to remove any sand.

3. Put the clams in a steamer over rapidly boiling water and cover the pan tightly.

4. Steam the clams until the shells open (about ten minutes).

5. Scoop the clams out of the shells and serve them with melted butter.

Shell collecting

It's fun to search the shore for special shells but instead of just collecting them, take a closer look at your "treasures." First of all, check to see if anybody's home. If there is an animal inside the shell, have a careful look at it, but let it go when you are finished. If the shell is empty it is safe to collect it.

Limpets
Chiton
Soft-shelled Clams
Ten-ridged Whelk
Periwinkle

Shelled animals are divided into three main groups: chitons, bivalves and univalves. Check out the shells on these pages to see what kinds you have found.

- Chitons are easily identified by their eight overlapping plates. These small, flattened shells are usually found attached to rocks.

- Bivalves, meaning two shells, include the familiar clams, mussels and scallops. Their shells are hinged together and attached by several muscles. On the beach, most bivalves have become separated. Look closely to find the place on the shell where it was hinged. Can you see smooth, roundish areas on the inside of the shell? These are scars that show where the muscles were attached.

- Univalves, such as snails, periwinkles and limpets, have only one shell. Their shell may be in a coil or spiral (snails and periwinkles) or cap-shaped (limpets).

Shell bells

Make some pretty wind chimes for your garden or balcony. You'll need:
- thin, strong cord (available in sewing stores)
- scissors
- a large darning needle
- a variety of different shells with small holes in them
- two Popsicle sticks
- craft glue

1. Cut four equal lengths of cord, about 50 cm each.

2. Using the needle, thread approximately the same number of shells along each cord, tying a knot below each shell so it can't slip off. If there is no hole, try piercing thin shells with the needle.

3. Make a cross with the two Popsicle sticks and glue them together in the centre.

4. Tie one cord of shells to each arm of the sticks. The shells should hang close enough so that they will bump each other when blown by the wind. If the cords are lopsided, add more shells to even out the weight of each side.

5. Tie a piece of cord to the centre of the sticks and hang your wind chimes outside where they will catch the wind.

Beachcombing

When you're looking for shells on the beach, you will likely come across a number of other plant and animal remains that have been washed up by the wind and the waves. Look for these common beach "finds" and others. Use a field guide (see page 63) to help you identify what you find.

SAND DOLLARS
Look for dried, bleached sand dollars. The round, hard shells of these animals are often used as decorations. Notice the five "petals" radiating from a central hole on the top. Turn the shell over to find the mouth hole in the centre.

CRAB SHELLS
Bits and pieces of crab shells often turn up in the sand. As crabs grow, they shed their old shell and grow a new one. The old shells are discarded and wash up on beaches.

SKATE EGG CASES
If you find a small black, leathery "pouch" with long, curled spines, you've found the egg case of a skate. Also known as a mermaid's purse, the pouch contained the eggs and developing young of a skate, a fish with a long tail and wing-like fins that lives close to the ocean floor.

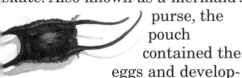

EYED FINGER-SPONGES
You may come across a piece of bleached spongy material with many finger-like branches and eye-like markings. Sponges look a bit like plants but they are really animals. If you have a hand lens, look closely at the pores covering the sponge. The sponge filters food out of the water that flows through the pores.

SEAWEED
Most seaweed grows attached to rocks underwater, but it often breaks off and is washed up on beaches. Try popping the balloon-like air bladders of Bladder Wrack (Rockweed) and Knotted Wrack, or wrap yourself in long ribbons of kelp. For more on seaweed, see page 36.

27

The seashore at

Pack a late-night snack and head to the shore on an evening when the tide is going out. Dress warmly, bring an adult and a flashlight and you're ready to explore the shore at night. Many animals come out in the dark because they are hidden from their enemies. Other creatures survive better in the cooler, damp night air.

When you're walking or sitting on the sand, you may notice some reddish-brown or bluish-gray, jumping creatures about the size of a jelly bean. These are sandhoppers, sometimes called beach fleas. If you're quick, you can catch one in your hand —

Sandhopper

they don't bite. Sandhoppers are tiny crustaceans, closely related to crabs and lobsters. During the day they burrow just under the sand in the upper beach to stay out of the hot sun. At night they venture out to feed on dead material left by the tide.

UP AND DOWN

The tiny floating plants and animals in the water, called plankton, move around at night. During the day, the plants float near the surface where the sun can reach them; at night they sink nearer the bottom. The animals move from the bottom to the top at night. Here they can feed more safely in the dark. Plankton is the most important life in the ocean because all other life depends on it. It is eaten by the tiny animals which are, in turn, eaten by larger and larger animals.

Plankton species

An underwater light

There's a lot happening in the shallow water at night, too. Tiny shrimps, small fish, crabs, clam worms and blood worms are just some of the creatures swimming around in the dark. You can get a good look at them by making this simple underwater light.

You'll need:
- a flashlight
- a jar with a tight-fitting lid, large enough to hold the flashlight
- a heavy rock
- waterproof tape
- some string

1. Turn the flashlight on and place it in the jar with the light pointing down. Add the rock to the jar as a weight.

2. Put the lid on tightly and seal it with some waterproof tape.

3. Tie string around the jar and carefully lower it into the shallow water. Hold on to the other end of the string. What do you see?

Rocky shores

What clings to the rocks along the seashore? Plenty! Discover how the plants and animals differ between the high tide and low tide zones. Find a mini ocean on the rocks and explore the amazing life of a tidal pool. Try cooking with seaweed. Spend your day hunting for fossils and then share your discoveries around an evening campfire at the shore. Don't forget the marshmallows!

On the rocks

Where can you go to see a great collection of plant and animal life, from water-logged seaweed and mussels to high and dry lichens and snails? Just look at a large rock or rocky cliff at low tide. Can you see the different colours, almost like layers, on the rocks? Like the beach, a large rock can be divided into zones.

High tide mark

Intertidal zone

32

Try to find where the high tide and low tide marks are. Between these marks is the intertidal zone. Take a hand lens and explore one of these rocks and discover the fascinating mini-world that clings to a rocky shore. Don't forget to check in the cracks and crevices, too.

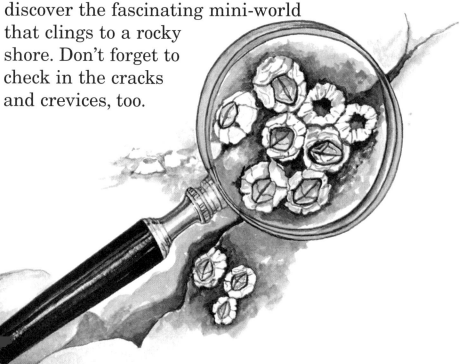

BARNACLES UP CLOSE

About half way between the high and low tide marks, you'll find colonies of hard white barnacles clinging to the rocks. Take a close look at a barnacle. Notice the series of shell-like plates that are tightly closed at low tide. When the tide comes in, these plates open up and the barnacle's feathery legs reach out to sweep plankton into the animal's mouth. When a barnacle colony gets crowded, the shells of the oldest barnacles grow longer, and younger barnacles grow up beneath them. Eventually the old barnacles are cast off the rock. See if you can find some old and young barnacles.

Exploring a rock

When the tide rolls out, pools of water are left behind creating mini oceans on the rocks. A visit to a tidal rock pool is like a close-up view of life under the sea. Each pool will hold a different assortment of plants and animals, and all of them will offer you a glimpse of how the plants and animals depend on each other for survival.

What to look for

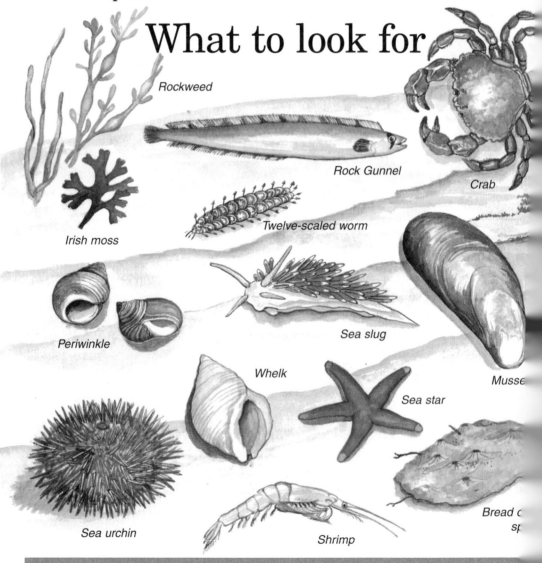

Rockweed, Rock Gunnel, Crab, Irish moss, Twelve-scaled worm, Periwinkle, Sea slug, Mussel, Whelk, Sea star, Sea urchin, Shrimp, Bread crumb sponge

34

pool

HIGH AND DRY?

When the tide is out, compare a tidal rock pool near the low tide mark with one near the high tide mark. Pools near the high tide mark are exposed to the hot sun for longer. Water evaporates more quickly from these pools, often leaving a white salt line around the rim of the pool. When water evaporates, the remaining water is saltier than the sea. The temperature of shallow pools can also rise very quickly. If you have a thermometer, compare the temperatures of the water in a rock pool near high tide and near low tide, to the ocean itself. The higher temperatures and saltiness in rock pools make it difficult for some creatures to survive.

Seaweed secrets

What do Sea Beard, Mermaid's Hair and Feather Bush have in common? They are all types of seaweeds. Seaweeds are large algae that grow in thick carpets along the coast.

They provide food and shelter for a huge variety of life including fish, periwinkles, sponges, sea cucumbers, and sea urchins. When you find a mass of seaweed, carefully check on and between the blades for the eggs and developing young of different marine animals. Notice the colour of the seaweed. There are three main groups: green, brown (including greenish-brown) and red (including purplish and nearly black). In general, you can find green seaweeds growing in the intertidal zone, brown ones growing near the low tide mark, and red seaweeds in the subtidal zone.

Seaweed for supper

Have you eaten any seaweed lately? If you've had commercial ice cream, pudding, frozen concentrates or chocolate milk, the answer is probably "yes." All of these things are made with carrageenin, a substance extracted from the seaweed called Irish Moss. Even your toothpaste might contain carrageenin! Here's another tasty way to eat seaweed. Irish Moss can be collected along rocky shores at low tide by raking.

The curly blades break off and the holdfasts remain on the rocks to grow new plants.

PARTS OF A SEAWEED

Unlike most land plants, seaweeds don't have roots, stems and leaves. Instead, they have a holdfast to attach them to a rock or other hard surface (A), a stipe which is similar to a leaf stalk (B), and a main body (C). When you examine a seaweed, try to find all its parts.

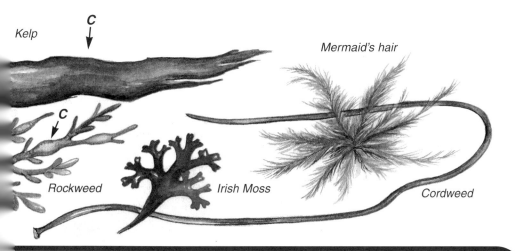

You'll need:
- 1/3 cup of Irish Moss
- 4 cups of milk
- 1/4 teaspoon of salt
- 1 1/2 teaspoons of vanilla
- a double boiler
- a bowl
- sugar
- a kitchen sieve
- toppings, such as sugar, cream, fresh fruit or flavoured sauces

1. Rinse the Irish Moss thoroughly under cold water to get rid of sand and dried salt.

2. Place the seaweed and the milk in the top of the double boiler and boil for about half an hour until the milk is slightly thickened.

3. Add the salt and vanilla and mix well.

4. Pour the contents of the pan through the sieve into a bowl and chill. When the mixture has cooled it will look a bit like white jelly. Serve it with your favourite topping for dessert.

Living under the

The most important life in the ocean is so small that you can't see it. It's called plankton. Plankton is made up of tiny floating plants and animals that all other life feed on. Without plankton, there would be no other life in the ocean. If you have a microscope, collect a sample of water in a jar and place a drop of it on a glass slide.

Look at the water through the microscope. You'll see beautifully shaped green organisms called diatoms, as well as some tiny, moving animals. The tiny animals feed on the diatoms. Insects and other small animals feed on the plankton animals. They are in turn eaten by small fish and other small animals. Larger fish and birds eat the small animals, and mammals eat the large fish. This sequence of eating and being eaten is called a food chain. Since most plants and animals are eaten by more than one thing, several food chains overlap and form a food web. Look at the simple ocean food web below to see how different plants and animals depend on each other for survival.

COLLECT SOME CLINGERS

If you look beneath a wharf or on rocks below the low tide level,

surface

you'll find a lot of different plants and animals clinging to these underwater surfaces. You can watch the amazing build-up of underwater plants and animals by "planting" a rock or weighted piece of wood in the water and checking it regularly over a week or two to see what has attached itself.

You'll need:
- a large rock and/or a weighted piece of wood
- a rope
- a magnifying glass
- a pencil and paper
- a camera (optional)

1. If you are using a piece of wood, tie a rock or other weight to it so it will sink in the water.

2. Tie a piece of strong rope to the rock or piece of wood and lower it into the water below the low tide line. Tie the other end of the rope to a wharf or other permanent anchor nearby, or to a floating buoy.

3. Every day, pull up your rock or piece of wood and check for life clinging to it. Use a magnifying glass to see the life close-up, especially the slimes that are the first to cling on. Make notes of what you see and, if you have a camera, take photographs.

4. At the end of your observations remove the rope and leave the rock or piece of wood in the water where the plants and animals can continue to grow.

Layers of life

Here's what you might find on your rock or piece of wood.

After a few hours: bacterial slime

After a day: plankton such as diatoms and protozoa

After two or three days: tiny animals such as hydroid and bryozoa.

After a week: larger animals such as barnacles and mussels, along with different algae.

Finding fossils

You can turn a walk on the shore into a journey into the past when you look for fossils — pieces of ancient history that record what life was like hundreds of millions of years ago.

Fossils are plant or animal remains that have been buried in sand or mud before they have rotted. Minerals from the sediment gradually fill all the air spaces in the bones, shells or woody parts, making them rock-like, or mineralized.

Some fossils found on Maritime shores are the first of their kind discovered in the world and they are very important to scientific study.

Nova Scotia is one of the best fossil finding places in the world. The Bay of Fundy shores, in particular, are well known as an excellent source of dinosaur footprints and bones and plant fossils.

The laws about fossil collecting are changing in the Maritimes. At this time, any fossil found on the shore belongs to the Province. It is strictly illegal to pull, or excavate anything out of the cliffs or reefs unless you have a research permit. Loose fossils on the beach may be picked up since they will likely be destroyed by the tides anyway, but we suggest that you take no more than two fossils so other people can make their own discoveries. We encourage you to learn about the fossils that you discover and try to find out what formed them. Visit the Fundy Geological Museum in Parrsboro, the Joggins Fossil Centre or one of the provincial museums and get some help from a fossil expert. If your fossil find is significant, it may be placed in the provincial collection to be shared by all.

Make your own fossil

You can see how fossils are formed with this quick and easy activity.
You'll need:
- a sponge
- scissors
- a bowl
- sand
- one cup of salt
- two cups of water

1. Cut the sponge into the shape of a bone, footprint, etc.

2. Half fill the bowl with sand and bury the sponge shape in it. The sand represents the layers of sediment that true fossils are buried in.

3. Mix the salt and water together and pour the mixture into the bowl. Salt is a mineral. Many other minerals are also present in the sediments that help form true fossils.

4. Leave the bowl in a sunny window for several days. When the sand looks dry, dig up your "fossil." It should be hard, like rock.
What happened?
   The sponge soaked the saltwater up into its air spaces. When the water evaporated, the salt was left behind, making the sponge mineralized, like a true fossil.

Campfires at the

If roasting hotdogs and marshmallows, singing songs and sitting out under the stars sounds like fun to you, then you'll love a campfire at the shore. Check the campfire tips and bring-along list below. Adult supervision is recommended.

CAMPFIRE TIPS

- build your campfire above the high tide mark. A ring of rocks around the fire pit will help contain the fire and keep young children from stepping into it.
- don't break branches off trees or shrubs for kindling.
- for cooking, keep the fire small since it will burn down more quickly, leaving hot coals perfect for roasting. Small fires are easier to put out at the end of the evening.
- when you are finished, let the fire burn right down. Smother the coals with water and sand until the fire is completely out.
- replace the rocks, remove all garbage and leave the shore as you found it.

shore

Bring-alongs and sing-alongs

- food
- matches
- kindling and firewood, if driftwood isn't around
- a flashlight
- a penknife for sharpening roasting sticks
- a garbage bag
- container for carrying water to put out the fire

Sing-alongs and stories

There's something magical about telling stories and singing around a campfire. The legends of Glooscap, the Mi'kmaq god from Blomidon on the Bay of Fundy, Nova Scotia, are excellent tales. Check a local library or bookstore for a copy of these stories. The Maritimes are also famous for their folksongs — try this oldtime favourite around your campfire.

Farewell to Nova Scotia, the sea-bound coast,
Let your mountains dark and dreary be,
For when I am far away on the briny ocean tossed,
Will you ever heave a sigh and a wish for me?

1. The sun was setting in the west,
   The birds were singing on ev'ry tree,
   All nature seemed inclined for rest,
   But still there was no rest for me.

2. I grieve to leave my native land,
   I grieve to leave my comrades all,
   And my aged parents whom I always held so dear,
   And the bonny, bonny lass that I do adore.

from *Folksongs of the Maritimes,* Kay Pottie and Vernon Ellis, Formac Publishing 1992.

Some shoreline animals up close

Find out how to identify the gulls that come to your picnic and use our handy checklist to keep track of the long-legged shorebirds you see. Have you ever wondered where all the jellyfish come from in the summer? Become a nature detective and find clues about the animals that live at the shore and in nearby saltwater marshes. If you're going on a ferry, check out our guide to some common offshore wildlife. And now that you've spent time at the seashore and learned about the plants and animals that live there, gather your family and friends together to help protect the seashore.

Wings over water

Few birds are as easy to recognize as seagulls. Whether they are following fishing boats at sea, dropping mussels on the rocks to break them open, or trying to share your picnic, seagulls seem to collect near food.

GULLS

The most common gulls along the East Coast are Herring Gulls and Great Black-backed Gulls. The Ring-billed Gull is also expanding its range into the Maritimes. All three adult gulls have white heads, necks, tails and underparts. Their webbed feet help gulls swim and their long beaks are great for picking up and eating fish, shellfish, large insects, and dead animals. When you see a gull, try to decide what kind it is using this simple guide to adult gulls.

TERNS

Have you seen a small, white gull-like bird swooping and diving into the water near the shore? It's probably a tern. Unlike the gulls, Common and Arctic Terns have forked tails, black caps, red bills and red feet and legs.

Ring-billed Gull
- yellow legs
- black ring on bill

Shorebirds

What has a long bill, long legs and is coloured to blend in with the sand and rock at the shore? If you said one of the twenty or so shorebirds found along the East Coast, then you're right.

Their long bills are adapted for probing into the mud to find worms and other small marine animals when the tide is out. Long legs help the birds to stay dry while walking at the edge of the water. In late summer, thousands of shorebirds flock to the mudflats on the upper Bay of Fundy and Minas Basin to fatten up before flying south for the winter. The best time to view shorebirds is when the tide is high or coming in. This brings the birds closer to shore for easier viewing.

Bird checklist

Grab your binoculars and use this handy checklist and a good field guide (see page 63) to help you identify some of the more common birds at the shore.

☐ Short-billed Dowitcher
Date_____ Place_____
☑ Dunlin
Date_____ Place_grass_
☑ Great Black-backed Gull
Date_____ Place_beach_
☑ Herring Gull
Date_____ Place_Kose_
☑ Ring-billed Gull
Date_____ Place_home_
☐ Great Blue Heron
Date_____ Place_____
☐ Red Knot
Date_____ Place_____
☑ Black-bellied Plover
Date_____ Place_~~St. Pete~~_
☑ Lesser Golden Plover
Date_____ Place_St. Pots_
☐ Semipalmated Plover
Date_____ Place_____
☐ Sanderling
Date_____ Place_____
☐ Least Sandpiper
Date_____ Place_____
☐ Semipalmated Sandpiper
Date_____ Place_____
☐ Spotted Sandpiper
Date_____ Place_____
☐ Arctic Tern
Date_____ Place_____
☐ Common Tern
Date_____ Place_____
☐ Ruddy Turnstone
Date_____ Place_____
☐ Whimbrel
Date_____ Place_____
☐ Greater Yellowlegs
Date_____ Place_____
☐ Lesser Yellowlegs
Date_____ Place_____

Jellyfish

Jellyfish get their name from the strong, jelly-like substance that forms their body. The sac-like body is shaped like an umbrella with a ring of tentacles around the edge. In the centre of the bottom is a mouth that leads to a large stomach. Around the mouth are "arms" that help put food, such as small fish, plankton and crustaceans, into the mouth. Jellyfish are in turn eaten by some fish and sea turtles.

ON THE MOVE
Jellyfish float on the open water and are carried a long way by the currents and the wind, but they can also swim. When a jellyfish wants to move, it contracts the muscles near the outer rim of the umbrella shape so the umbrella looks like it is closing up. When it relaxes the muscles, water rushes into the animal's stomach until it is full. Now the jellyfish quickly contracts the muscles again and forces the water out with such power that the jellyfish shoots forward, like a rocket.

A DOUBLE LIFE
The life cycle of a jellyfish includes a non-swimming stage when it is called a polyp and lives attached to a rock, wharf or other solid object underwater. Later it becomes a free-swimming umbrella-shaped jellyfish that you see in the water. In early spring thousands of tiny jellyfish break off from the colonies of polyps and spend the summer drifting with the currents. By mid-summer they have grown into large jellyfish that may appear in huge numbers near the shore. During August and September the jellyfish lay their eggs and the young hatch and attach themselves underwater to spend winter in the

non-swimming stage. Soon after egg-laying, the adult jellyfish die.

STRINGS OF STINGS

The long tentacles of a jellyfish contain stinging cells that are used to capture food and for self-defence. Stinging cells contain

poison to paralyse prey, and some can shoot out masses of tiny threads to tangle prey. Both types of jellyfish commonly found off the East Coast in late summer and early fall — Moon Jellyfish and Lion's Mane or Red Jellyfish — should be avoided, but the sting of the Lion's Mane is the most severe. If you happen to find a jellyfish washed up on the beach, do not touch it because its tentacles can still sting several hours after the animal has died.

Home-made remedies for jellyfish stings

If stung, try one of these home-made remedies to help relieve the pain and rash.

- Put vinegar or a cola drink on the sting to relieve pain.

- Make a herbal tea by adding a handful of leaves and flowers from St. John's Wort (a common roadside wildflower) and Stinging Nettle (use gloves to pick this) to two cups of boiling water. Soak a cloth in the water and bathe the stings. You can make the tea ahead of time and carry it to the beach in a thermos.

- A cloth soaked in hot water can help relieve the pain of a sting.

What a crab!

When you grow bigger, you grow out of your clothes and need new ones. But imagine if you also had to get a new house! That's what hermit crabs have to do. Instead of just living in their own shell like most crabs, hermit crabs live inside the abandoned shell of a snail.

When the crabs grow, their house gets too small and they have to find a bigger snail shell to call home. Look for hermit crabs scurrying over the sand and mud near the water line or along the bottom in shallow water. You may also find them in tide pools, especially near the base of large rocks. When they are frightened, they pull their legs into the shell and cover the opening with the large pincers attached to their front legs.

EMPTY SHELLS

Along the beach you may find the empty shells of some other common crabs — Rock Crabs and Jonah Crabs. As these crabs grow, their hard shells split along the back where the body and tail are joined and the crab crawls out. This is called moulting. Beneath the old shell, the crab grows a new one that hardens in a few weeks. Before the new shell hardens, crabs have no protection from predators so they stay hidden in seaweed to avoid hungry fish, herons, gulls and diving ducks. To find young crabs, check under pieces of wood or rocks on sandy and rocky shores and in tide pools. Adults are often partly buried in the sand, with their large pincers near the surface ready to grasp worms, mussels, sea urchins or dead fish.

Use a field guide to identify these other common crabs: Green Crab, Lady Crab, Lesser Toad Crab.

Hold a crab race

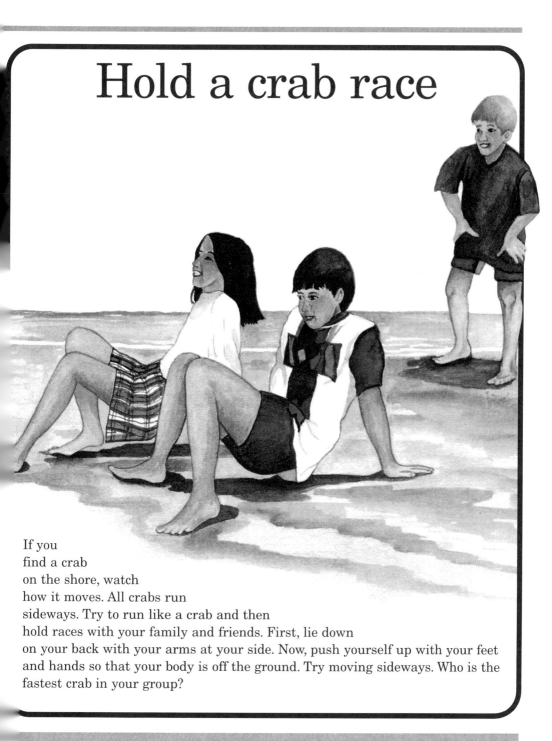

If you
find a crab
on the shore, watch
how it moves. All crabs run
sideways. Try to run like a crab and then
hold races with your family and friends. First, lie down
on your back with your arms at your side. Now, push yourself up with your feet
and hands so that your body is off the ground. Try moving sideways. Who is the
fastest crab in your group?

Saltmarshes

In sheltered bays and other shoreline areas that are protected from rough waves, you may find saltwater marshes. These rich ecosystems provide food and shelter for an amazing assortment of plants and animals.

A number of flowering plants including grasses, sedges and rushes grow in saltwater marshes. Their roots trap sediment from the water and bind it together so that it cannot be washed away by the tides. Many of these plants are specially adapted to the salty water. They have thick, fleshy stems or leaves that store fresh water and can get rid of the salt that is absorbed through their roots.

MARSH ZONES

Like other shoreline areas, saltwater marshes are divided into zones — high marsh and low marsh — depending on how much they are flooded during tides. Plants typical of the high marsh include Marsh Hay, Sea Lavender, Seaside Plantain, Milkwort, Arrow Grass and a variety of grasses, rushes and sedges. The low marsh is home to cordgrass, some algae, Samphire, Sea-blite, Orach and Seaside Sand Spurry. Many of these plants are eaten by ducks and geese, and they provide shelter and food for a variety of crabs, insects, spiders, Soft-shelled Clams, snails and Ribbed Mussels. Muskrats also feed on the stems and roots of marsh plants and use the plants to make their dome-shaped homes in the marsh.

Crafty plants

Two saltmarsh plants are favourites for crafts. Sweetgrass, which grows at the upper edge of the high marsh, has a beautiful fragrance and is traditionally used in basket-making and other crafts. Sea lavender also grows in the high marsh. Its dense colonies of small, bluish flowers can be collected in the fall and hung upside down to dry. They are lovely additions to dried flower arrangements, wreaths and other decorations.

Become a nature

When you go for a walk at the shore, you may not see many of the animals that live there or visit at certain times of the day or night. But you can discover what is around by being a nature detective and looking for the clues they leave.

For instance, raccoons, skunks and mink may visit the water's edge at night for food — clams, mussels, worms and other creatures that are left behind when the tide goes out. Watch for their tracks in the wet sand before the tide comes in.

Muskrats may live nearby in a shallow saltwater marsh. They can be identified by tracks, but even more easily, by their dome-shaped houses made out of dead marsh vegetation and mud.

Many birds feed and nest at the shore. Look for the large nests of sticks built by Osprey, often in dead trees near the shore. Great Blue Herons also build nests of sticks and grass in the tops of

detective

large trees. Since these birds nest in colonies, you will see several nests in a single tree. Bank Swallows and Belted Kingfishers prefer nesting in holes near the tops of sandy cliffs. Spotted Sandpiper nests are simply depressions on the ground lined with leaves and grass. They are usually hidden in the grass or under a shrub in the upper beach area and are active in late May and June. All bird nests should be watched from a safe distance so the birds are not disturbed. Since many birds return to the same nest year after year, nests should not be removed or collected.

Animals ahoy!

If you take a ferry trip from shore to shore in the Maritimes, or you sit on a rocky shore and watch out over the water, you may see the sleek, bobbing head of a seal or a spout from one of the magnificent whales that live in the ocean. Observing whales, seals, porpoises, dolphins and seabirds is a great hobby for boaters and landlubbers alike.

Minke

Fin Whale

Humpback Whale

Se...

Harbour Porpoise

Atlantic White-sided Dolphin

WHALE WATCHING

Whales are most commonly seen in the spring and fall as they pass by our shores during migration between their summer feeding grounds in the north and their winter birthing grounds in the south. Look for their spouts, dorsal fins, and tail fins as they dive. Large flocks of seabirds feeding out at sea may be a clue to nearby whales who are also feeding on the fish. Some special whale watching sites in Nova Scotia include Brier Island, the west coast of Cape Breton Island, Scot's Bay, St. George's Bay, and Halifax Harbour. The Fundy shore and Grand Manan in New Brunswick are also good spots, along with the Atlantic coasts of all the Maritime provinces. Whales may also be sighted from any of the ferries running between New England, Nova Scotia, New Brunswick, Prince Edward Island and Newfoundland.

Use this checklist of some common wildlife seen from the shore and a field guide to help you with identification.

Best times for viewing

Animal	Spring	Summer	Fall	Winter
Atlantic White-sided Dolphin		X	X	
Common Dolphin		X	X	
Harbour Porpoise		X	X	
Grey Seal		X	X	
Harbour Seal		X		
Harp Seal		X		
Hooded Seal	X			
Atlantic Pilot Whale	X		X	
Fin Whale			X	
Humpback Whale	X	X	X	
Minke Whale	X	X	X	
Double-crested Cormorant	X		X	
Great Cormorant	X	X	X	
American Black Duck	X	X	X	
Common Eider		X	X	
Canada Goose	X	X		
Common Loon	X		X	X
Red-throated Loon		X	X	some areas
Red-breasted Merganser	X		X	X
Common Merganser		X		
Black Scoter		X	X	
Surf Scoter	X	X	X	
White-winged Scoter	X	X		

S.O.S. Save our

Throughout this book you have discovered the amazing variety of plants and animals that call the seashore home. These plants and animals depend on each other for survival, but none can survive without their habitat

Whether it's a rocky shore, sandy beach, mudflat or saltmarsh, the destruction of a habitat leads to the loss of the species that live there. Some of the major threats to shoreline species include pollution from oil spills, toxic chemicals and garbage, and motor vehicles. Several endangered species, including the Piping Plover, Leatherback Turtle and Northern Right Whale depend on healthy shorelines and waters for their survival. You can help protect the shoreline and its wildlife by following the Conservation Tips on page 7. Spread the word to your family and friends that the shoreline is a great place to visit and tell

Leatherback turtle

them how they can help protect it, too. You can also volunteer your time or money to help organizations that are working to protect the shoreline and its species. Contact the local naturalists' group or the Clean Nova Scotia Foundation for information on groups that are involved in shoreline conservation work.

horelines

Right whale

PIPING PLOVER GUARDIAN PROGRAM

The Piping Plover is a small shorebird that nests in spring on white sandy beaches, close to the dunes. Its declining population is endangered mainly because of human disturbance, particularly from ATVs and other vehicles on the beaches. Garbage left on beaches may also attract predators such as raccoons, skunks, crows, ravens and gulls. Naturalist groups, government agencies and volunteers are protecting the nesting sites of these birds in Nova Scotia, New Brunswick and Prince Edward Island through the Piping Plover Guardian Program. Signs are put up on beaches where Piping Plovers are known to nest, helping people understand the importance of avoiding these areas. Volunteer guardians also visit the sites to help explain the program to people on the beach. Help protect Piping Plovers by following these simple guidelines:

1. Keep your beach activity below the high tide line and dune grass area on white sandy beaches.

2. Keep pets leashed at all times on beaches.

3. Help keep the beaches free from litter that might attract predators.

4. Don't approach a Piping Plover nest, young or adult bird.

5. Become a volunteer Guardian by contacting the Canadian Wildlife Service or your provincial wildlife division of the government.

Piping plover

East Coast National, Provincial and State Parks in Canada and New England

(C) indicates camping facilities

When your family is travelling along the east coast in Canada and New England, check out some of these national, provincial or state parks. All of them are located along the shore where you can explore their beaches and try out some of the activities in this book.

Newfoundland and Labrador
Information:
1-800-563-6353

National Parks:
Gros Morne National Park (C)
Terra Nova National Park (C)

Provincial Parks:
Backside Pond (C)
Bellevue Beach (C)
Black Bank
Blow Me Down (C)
Bottle Cove
Cape Ray (C)
Chance Cove (C)
Codroy (C)
Deadman's Bay
Dildo Run (C)
Frenchman's Cove (C)
Grand
Indian Cove Neck
Indian River
J.T. Cheeseman (C)
Jack's Pond (C)
La Manche (C)
Maberly
Mummichog (C)
Northern Bay Sands (C)
Notre Dame
Piccadilly Head (C)
Pinware River (C)
Pistolet Bay (C)
Point au Mal
Point LaHaye
Salmon Cove Sands
Sandbanks (C)
Sop's Arm (C)
The Arches
Windmill Bight (C)

New Brunswick
Information:
1-800-565-0123

National Parks:
Fundy National Park (C)
Kouchibouguac National Park (C)

Provincial Parks:
The Anchorage (C)
Castalia (C)
Escuminac
Herring Cove (C)
Maisonette
Murray Beach (C)
New River Beach (C)
Oak Bay (C)
Parlee Beach (C)
Pokeshaw (C)
Val Comeau (C)

Prince Edward Island
Information:
1-800-565-0267

National Parks:
Prince Edward Island National Park (C)

Provincial Parks:
Argyle Shore
Belmont (C)
Brudenell River (C)
Cabot Beach (C)
Campbell Cove (C)
Cedar Dunes (C)
Chelton Beach
Fisherman's Haven
Green Park (C)
Jacques Cartier (C)
Linkletter (C)
Lord Selkirk (C)
Northumberland (C
Panmure Island (C)
Pinette
Red Point (C)
Souris Beach
Union Corner
Victoria
Wood Islands

Nova Scotia
Information:
1-800-565-0000

National Parks:
Cape Breton Highlands National Park (C)

Provincial Parks:
Amherst Shore (C)
Anthony
Battery (C)
Bayfield Beach
Bayswater Beach
Blomidon (C)
Cabot's Landing
Caribou (C)

62

Suggested Field Guides

lam Harbour Beach
leveland Beach
rystal Crescent Beach
ominion Beach
ive Islands (C)
raves Island (C)
ulf Shore
eather Beach
awrencetown Beach
ennox Passage
artinique Beach
avilette Beach
elmerby Beach
orthport Beach
oint Michaud
omquet Beach
ondville Beach
ort Maitland Beach
owell's Point
ueensland Beach
ainbow Haven
isser's Beach (C)
ushton's Beach
and Hills Beach
econd Peninsula
ummerville Beach
aylor Head
he Islands (C)
idnish Dock
or Bay
aterside Beach

EW ENGLAND

ll parks listed have
mping facilities)

onnecticut

ate Parks:
ammonasset Beach
State Park

Rocky Neck State Park

Maine

National Parks:
Acadia National Park

State Parks:
Lamoine State Park
Warren Island State Park
Cobscook Bay State Park

Massachusetts

State Parks:
Boston Harbor Island
   State Park
Salisbury Beach State
   Reservation
Scusset Beach State
   Reservation

New Hampshire

State Parks:
Hampton Beach State
   Park

Rhode Island

National Parks:
Fishermen's Memorial

State Parks:
Fort Getty Camping and
   Recreation Area
Burlingame State Park
Charlestown Breachway
   State Campground

Field guides can help you identify some of the things you will find during your seashore adventures. Look for these books in your local library or bookstore.

A Field Guide to the Birds of North America East of the Rockies, by R.T. Peterson. Houghton Mifflin Company, Boston, 1984.

A Field Guide to Wildflowers, Northeastern/ Northcentral North America, R.T. Peterson and M. McKenny. Houghton Mifflin Company, Boston, 1968.

Fundy Shore Ecotour Travel Guide published by Central Nova Tourist Association, Halifax.

Seashores. A guide to Animals and Plants along the Beaches, by H.S. Zim and L. Ingle. Golden Press, New York, 1989.

Seashores. Summer Nature Notes for Nova Scotians, by Merritt Gibson. Lancelot Press, Hanstport, Nova Scotia, 1987.

The Fossil Cliffs of Joggins by Laing Ferguson. Nova Scotia Museum, Halifax, 1988.

Tidal Life. A Natural History of the Bay of Fundy, by Harry Thurston. Camden House Publishing, Ontario, 1990.

Index

barnacles 6, 33, 39
beachcombing 26–27
beach formation 8
beach sweep 7
birds 20–21, 46–49, 56–57, 60–61
  Bank Swallows 57
  Belted Kingfishers 57
  Dunlin 21
  Great Blue Heron 56–57
  gulls 20, 46–47, 61
  Osprey 56–57
  Piping Plover 60–61
  shorebirds 20, 48–49, 61
  Spotted Sandpiper 20, 57
  terns 46–47

campfires 42–43
clams 21–23, 56
  see also shells
conservation 7, 60–61
crabs 29, 34, 52–53, 55
  see also shells
crafts 19, 25, 55

dunes 7, 10–11

endangered species 60–61

field guides 63
fish 29, 36, 46, 50, 52, 59
  flounder 20
  Rock Gunnel 34
food chain 38
fossils 7, 40, 41

jellyfish 50–51
  polyps 50
  stings 51

Leatherback turtle 60

marshes 54–56
mink 56
muskrat 55–56
mussels 24, 34, 46, 55, 56

nests 7, 56–57, 61
night 28–29, 42–43, 56

parks 62–63
plankton 28, 33, 38–39, 50

raccoons 56, 61
rock pool 34–35

saltmarsh see marshes
sandhoppers 28
seal 58–59
sea slug 34
sea star 34
sea urchin 34, 36, 52
seaweed 27, 34, 36–37
  Cordweed 37
  Feather bush 36
  Irish moss 34, 36–37
  kelp 27, 36–37
  Knotted Wrack 27
  Mermaid's Hair 36–37
  Rockweed 27, 34, 36–37
  Sea Beard 36
shells 24–25
  bivalves 24
  Clams, Soft-shelled 21, 24, 55
  chitons 24
  crab 26, 34, 52–53
  limpets 21, 24
  periwinkles 24, 34, 36
  Sand Dollars 20, 26
  whelks 24, 32

univalves 24
shrimp 29, 34
skunk 56, 61
sponges 27, 34, 36
  Eyed Finger-sponge 2
  Bread Crumb sponge 34
super scooper 14, 22

tides 6–9, 18, 32–33, 35
  tide tables 6–7
  tide schedule 7, 9
tracks 20–21

underwater
  exploration 38
  light 29
  viewer 15

whales 58–59
  Atlantic White-sided Dolphin 59
  Fin Whale 58–59
  Harbour Porpoise 59
  Humpback Whale 58–59
  Minke 58–59
  Northern Right Whale 60–61
whelks
  see shells
worms 21–22, 34, 48, 52, 56
  bloodworms 22, 29
  clam worms 22, 29
  Lugworm 21
  Twelve-scaled worm 34

zones
  beach 10–11
  rock 32
  marsh 55